Bibliographic information published by the German National Library:

The German National Library lists this publication in the National Bibliography; detailed bibliographic data are available on the Internet at http://dnb.dnb.de .

Imprint:

Copyright © 2015 GRIN Verlag, Open Publishing GmbH
Print and binding: Books on Demand GmbH, Norderstedt Germany
ISBN: 9783668454460

This book at GRIN:

http://www.grin.com/en/e-book/365411/aquaculture-profile-of-the-naitasiri-province

Epi Batibasaga

Aquaculture Profile of the Naitasiri Province

Naitasiri Aquaculture

GRIN Publishing

Aquaculture Profile of the Naitasiri Province

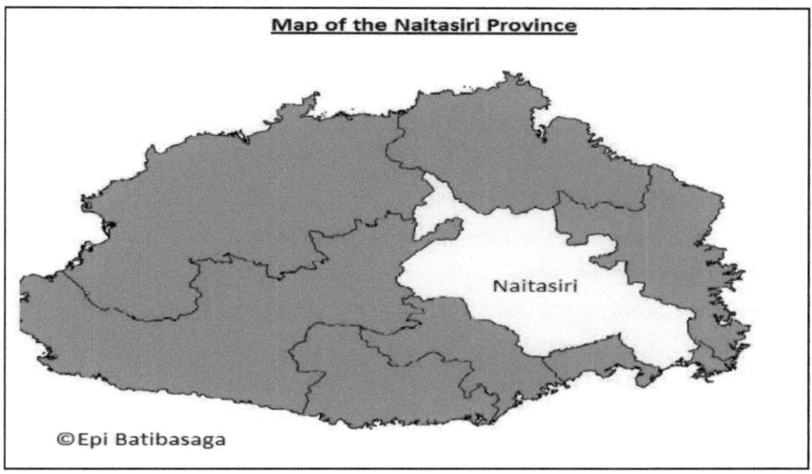

Figure 1. Naitasiri Province

Abstract

Aquaculture has existed in Fiji for over 60 years and has been beneficial to many through the associated economic and social benefits it has in offer. In the Fiji inland communities of the Naitasiri province the potential of freshwater aquaculture is slowly being unlocked and farmers are now more than ever eager to move the farming to the next level. The abundant water sources, suitable land and the overwhelming reliance on farmed freshwater species for food and income is one that should be fully capitalized but in order to do so more research needs to be carried out to support such claims and better understand what is involved. Developing an inventory on aquaculture farming in the Naitasiri province will allow us to understand the current status, the farming capacity, mechanisms involved, systems and processes so that weaknesses and threats be brought to light and the non-value adding components removed thus to recommend ways to improve farming for the Naitasiri province and for Fiji freshwater aquaculture.

Keywords: Freshwater Aquaculture, Naitasiri Province, Inventory, Aquaculture Profile

Table of Contents

Acknowledgements

I would like to thank the following for without whom; the compilation of this report would not have been possible:

1.	Ms. Salote Waqairatu of the University of the South Pacific for the guidance, mentorship and supervision throughout this research.

2.	The Fisheries Department for allowing me to be part of the research, funding and providing the much needed human resource

## 1.0	Introduction

The earliest known history of aquaculture occurred around the 2000-1000 B.C in China where Hickling an English aquaculture author cited S.Y Lin's "a noted Chinese aquaculturist" at the time the common carp was farmed (FAO, 2015). Written history of aquaculture can be dated back as early as 473B.C where 'Fan Li' wrote the book "The Classic of Aquaculture" (FAO, 2015). This "first record describe the structure of ponds, the method of propagation of the common carp and the growth of fry (FAO, 2015). The term '**Aquaculture**' refers to the breeding, rearing and harvesting of animals and plants in all types of aquatic environments including rivers, oceans, lakes and ponds (NOAA, 2015). Almost half the seafood consumed globally is farmed (Aquaculture) due to many wild fisheries not meeting the demand of the ever increasing human population (NOAA, 2014). This type of farming is many times commonly recognized for producing food organisms like prawns, shellfish and fish (The State of Queensland, 2010-2015). According to (SeaWeb, 2004) "Aquaculture is the fastest growing sector of the world food economy, increasing by more than 10% per year and currently accounts for more than 30% of all fish consumed". Aquaculture farming is further divided into two main categories and these include freshwater and marine (The State of Queensland, 2010-2015). Over 80 percent of the world production is linked with Asia (Meade, 1989). In 2011, contribution to world production by inland aquaculture was a staggering 44.3 million tonnes which is 69.7 percent of all aquaculture production (FAO, 2012). For the same year both farmed species namely; tilapia and prawn production estimates in Fiji equalled 218.51 tonnes, harvested by 265 farmers in 531 ponds (Ministry of Fisheries and Forests, 2011). Estimated Aquaculture revenue earned in 2013 totalled $550,000 (Ministry of Fisheries and Forests, 2013).

An article published in the FAO website states, "per capita supply from aquaculture increased from 0.7 kg in 1970 to 7.8 kg in 2006" (FAO, 2015). The Asia Pacific region contributes greatly to world aquaculture by being the highest producer "which accounts for 89 percent of the production" (FAO, 2015). Farming in freshwater or 'Freshwater Aquaculture' for Fiji began in 1954 with the introduction of Tilapia at Nacocolevu Agriculture Research Station in Sigatoka to be used as pig feed (SPC, 2011). Carps, Prawns, Black Bass and Ornamental Fish were introduced later on with the recognition of their value, following overseas training and research and the huge potential it holds for Fiji. National aspirations were to develop aquaculture in rural areas which could be an alternative to limited inshore fisheries resources (SPC, 2011). Another goal was to develop aquaculture sustainably to support food security, for poverty alleviation, employment and foreign exchange earnings. The Aquaculture program in the rural areas was started in 1982 with 27 farms setup to supplement protein requirements. The development strategy then moved to increasing farming capacity which saw the shift from monoculture to poly-culture then to integrated farming practises. The leading stakeholder for aquaculture farming in Fiji is the Ministry of Fisheries and Forests which is responsible for coordination, policy formulation, dispute resolution and technical services (SPC, 2005). Others include Ministry of Lands, Ministry of Agriculture, Ministry of Health, Ministry of Regional development, Ministry of Fijian Affairs, Ministry of Public Enterprises, Ministry of Education, Ministry of Women Affairs, Tilapia Association, SPC, USP, World Fish, FAO, JICA and ACIAR. There have been big plans for the development of the commodity and this was highlighted in the 2005-2010 Fiji Islands Freshwater aquaculture sector plan where 2010 expected production were to be 6840 metric tonnes valued at around 62 million dollars but this did not eventuate.

Aquaculture farming has not always been an easy venture and growth for Fiji has slowed over the years opposite of what was expected of it. Developments have been hindered by the stringent environment, often governed by many problematic constraints (James Teri, 2007). One of these is the availability of very little data with regards to freshwater aquaculture in the rural areas as was in the early days (Foscarini, 1990). Commercial and subsistence scale of operations differ due to many reasons where in rural areas these are mostly controlled by socio-economic factors, the commercial is affected by wider economic and political conditions (James Teri and Tim Pickering, 2004). Other setbacks to farming were due to problems with fry production, feed supply, man power, land tenure, feeding and financial constraints. For Aquaculture the collection of basic data on farmers, harvests, prices, intensity, marketing to name a few is essential for policy, planning and management of Aquaculture (FAO, 2015). The collection, processing and analysing and redistributing data has been the Food Agriculture Organization's major roles as well as to promote improved techniques and approaches to data collection (FAO, 2015). For member countries this responsibility has been placed on the government office in charge of Fisheries, the Fiji Fisheries Department.

Data quality during collection impacts directly on the analyses made on them, meaning if low quality data were to be collected then low quality results were to be expected (FAO, 2015). World, Regional data collection systems are influenced by national systems making data a very vital information base and one that should not be taken lightly. Also data collection have had problems of its own like the inadequate human resource available to collect data, budget limitations, infrequent surveys and the availability of properly trained individuals to conduct the surveys (FAO, 2015). Yearly publications on world fisheries made by the Food Agriculture Organisation show more generalized information where only production in tonnes is noticeable for Aquaculture. This study will attempt to bring to light details about the unsung heroes in the farming communities who contribute to the existence of the data, try to capture the reality in the farming and missed information that could be made useful for proper management of aquaculture in Fiji. Naitasiri Province being one of the Freshwater Aquaculture Farming hotspots is one of the eight provinces on the main island of Viti Levu with boundaries beginning from the Suva Peninsular extending inland and terminating just at the base of Fiji's highest peak known as Mount Tomaniivi. The province covers an area of 1,666 square kilometres which is approximately 643.25 square miles; there is suitable land for the expansion of Aquaculture. It is characterized as being an ideal study area, due to the fact that it houses the second largest concentration of Freshwater Aquaculture farmers in Fiji with a total of 59 farmers in 2011 and second only to the Tailevu Province which had 89 farmers for the same year.

Great Interests in aquaculture in the Naitasiri province can only be acknowledged by observing the small pressure put on inshore fisheries irrelevant of the limited available inshore fishing grounds for Naitasiri. In 2014 a total of 38 Fishing Licences were issued to the Naitasiri Province, 15 of which fish for marine fauna with the remaining 23 fishing in freshwater systems findings of which are based upon the results of a personal interview conducted with Mr. Maikali Rasuaki of the Fiji Fisheries Departments' Monitoring Control and Surveillance unit in the Central Divisional office (Rasuwaki, 2015). Also according to Mrs. Nandita Naidu of the Fiji Fisheries Department, geographically the province of Naitasiri in contrast to the Tailevu province rely primarily on freshwater fish due to most of the provinces' fishing grounds being located on rivers and adjoining watersheds, therefore it is essential to promote aquaculture in the province to meet protein dietary needs, improve financial status and support dependents on this type of venture for their livelihoods (Naidu, 2015). The freshwater Aquaculture farming in the province is continuously being supported by the Fiji Government initiatives so that locals or the landowners utilize idle land to benefit them (The Fiji Times Online, 2015) , ensure the importance of meeting needs for human sustenance and providing a supplement of animal protein to the protein deficient rural inland communities (FAO, 2015). The urgency and the dire need to carry out such a survey is because there has not been any survey on the status of Aquaculture in the province since 2013 and reports on this to date is still unpublished.

1.1   The Study will focus on the following objectives:
<u>Broad Objectives</u>

•   Assess the status of Freshwater Aquaculture in the Naitasiri Province in Fiji.
<u>Specific Objectives</u>

•   Develop an inventory on Aquaculture farming that provides precise, realistic and on the ground information which may be used as a baseline for the development and effective management of aquaculture in the Naitasiri province and for Fiji.
•   Identify Trends and make recommendations for the betterment of farming in the province.

2.0   Materials and methods

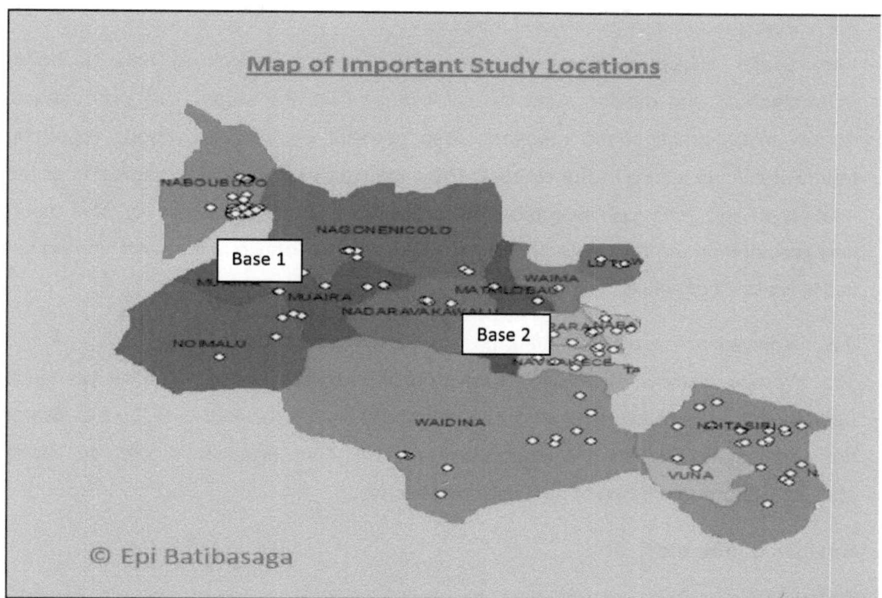

Figure 2. Important Study Locations

2.1   Logistics

Two 4x4 vehicles will be used to assist researchers in reaching the study sites. Two base camps were set up in the Viria and Navuakece districts to allow for easy and timely access to the upper regions and the lower regions of the Naitasiri province.

2.2   G.P.S locations

These locations were recorded using handheld G.P.S devices upon visitation of farms. Coordinates recorded were used to locate ponds (farms) and to map effectively.

2.3   Survey Questionnaires

Survey questionnaires were used to gather useful data on the status of freshwater aquaculture from individual farmers in the Naitasiri province. The participatory approach was utilized in gathering information so to allow for maximum output, create an environment for farmers to freely express their views and experiences in farming and to ensure that the results provide a better insight into the reality of the farming in the Naitasiri province. A total of 224 Farmers were interviewed.

2.4 Published and Unpublished Literature

Very useful published literature was used in this research to help in better understanding and relating some of the findings from the study. This was used to narrow focus areas in the research. Also relevant unpublished reports regarding freshwater aquaculture in the Naitasiri Province was useful in the compilation of the final report and getting an insight on farmers and the farming status in Fiji. This report was provided by the Fisheries department and included previous research carried out in the Naitasiri province.

2.5 Analysis on proximity to the urban centre (Nausori Town)

The average distance to Nausori town from the various farming districts would be used to study the relationship and trends that may be associated with the freshwater aquaculture farming in the Naitasiri province. This analysis would be made specifically for the context of the Naitasiri province.

Species of interest:

All tilapia species farmed in Fiji some of which include; (Oreochromis *Niloticus*, Oreochromis Mossambicus) and all Prawn species farmed in Fiji.

2.6 Photographs from the survey

Photographs would be taken to assist in the identification of certain sites and to give a visual idea of what farms look like. There will be two cameras for the two separate teams.

2.7 Fisheries License records

License records will be provided by the Department of Fisheries to give us a fair idea about the pressure on the fishing grounds near the study area, dependency on aquatic resources and the impetus of having aquaculture.

3.0 Results

3.1 Global Aquaculture Production

Year	1990	1995	2000	2005	2010	2012
Production in tonnes	13,074,679	24,382,522	32,417,781	44,297,145	59,037,416	66,633,253

Table 1. World Production of Food fish from Inland Aquaculture and Mariculture

	Inland Aquaculture 2012 Production(tonnes)
Fish	38,599
Crustacean	2,530
Mollusc	287
Other Species	530
Total	**41,946**

Table 2. World 2012 Production of Finfish, Crustaceans, Mollusc and Other Species

3.2 National Aquaculture Production

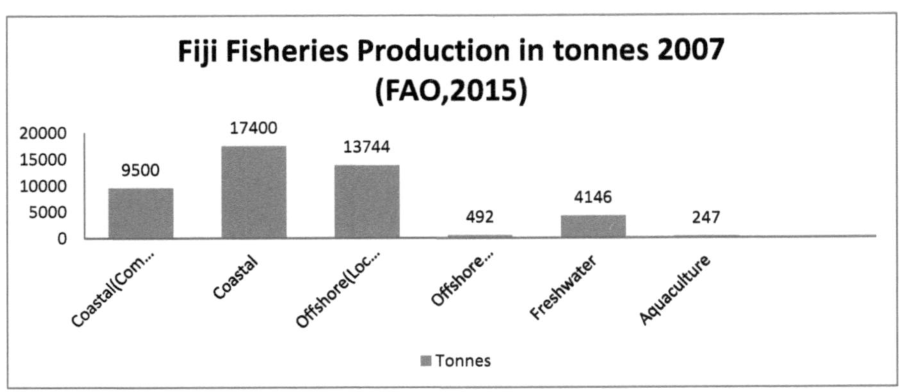

Figure 3. Fiji 2007 Aquaculture Production

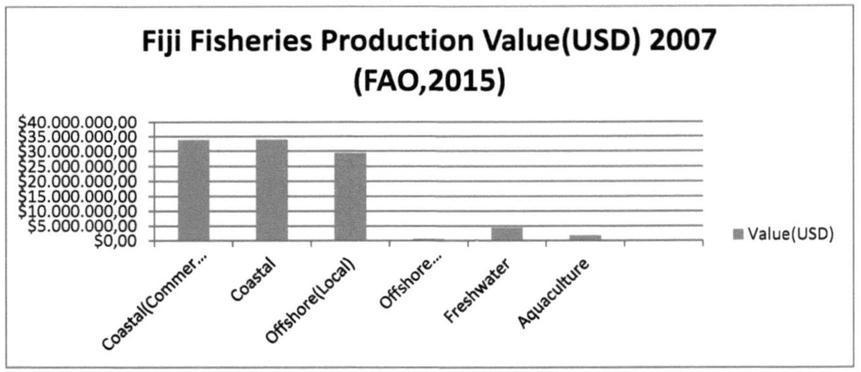

Figure 4. Year 2007 Aquaculture Value

Year	Country	Tonnes	Percent
2012	Republic of Fiji	200	0.10%

Table 3. Fiji 2012 Aquaculture Production

3.3   Fiji Fisheries Department Inshore License Records

2014 Fisheries Department Fishing Licence Records

- Northern Division – 815 issued

- Central – 331

- Eastern – 142

- Western - 102

3.4   Naitasiri Province Fishing Licences issued in 2014

Naitasiri Province

- Marine – 15

- Freshwater - 23

3.5   Survey Results

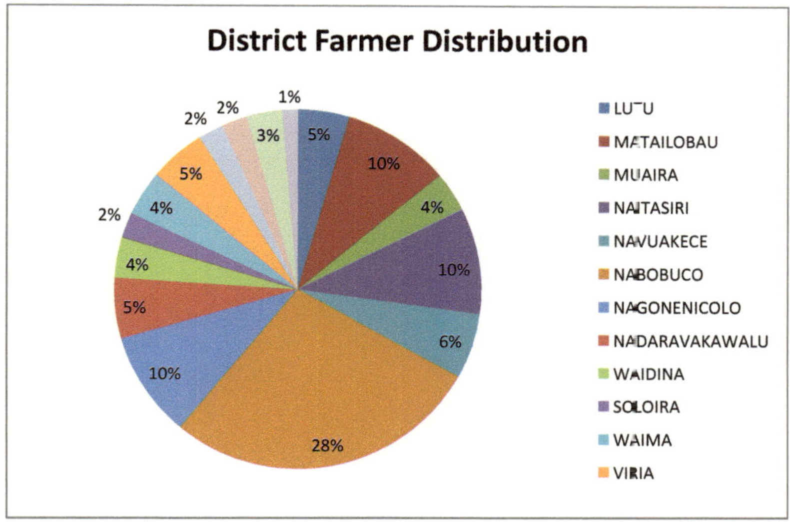

Figure 5. Farmer Distribution by District

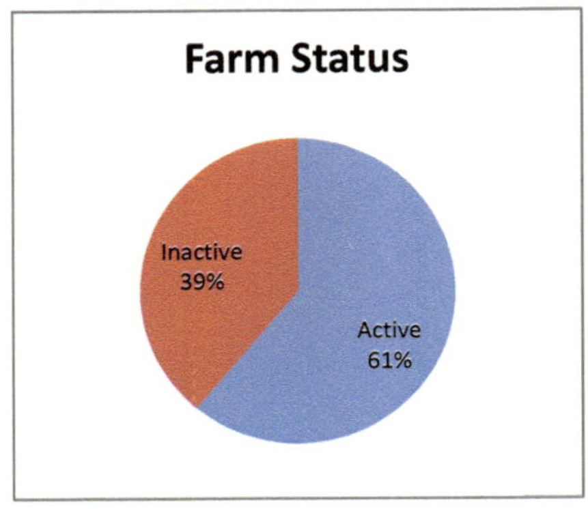

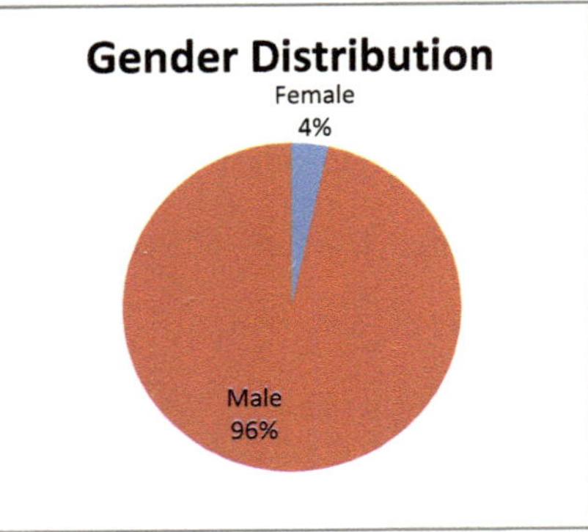

Figure 6. Farm Status

Figure 7. Gender Distribution

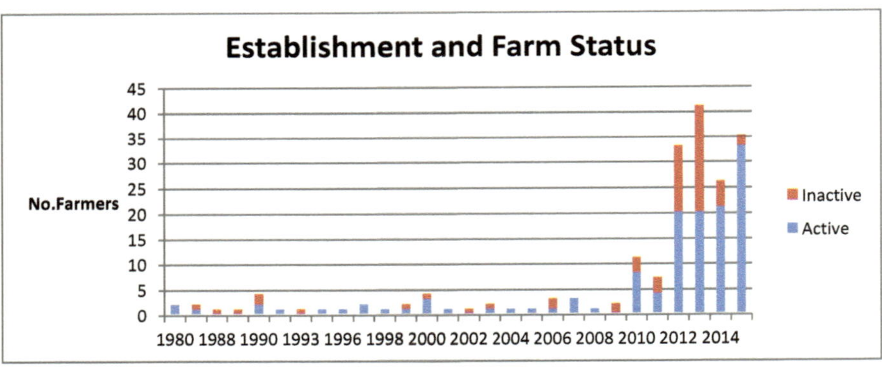

Figure 8. Farm Status 1980-2014

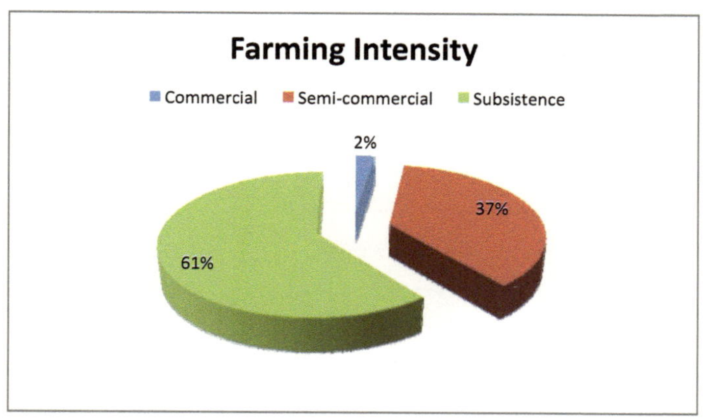

Figure 9. Naitasiri Faming Intensity

Correlation Matrix

	Age	Cost of Pond Construction	Total Pond Area(m2)	No: of Ponds	Income
Age	1.000				
Cost of Pond Construction	.114	1.000			
Total Pond Area(m2)	.066	.707	1.000		
No: of Ponds	-.132	.013	.018	1.000	
Income	-.001	.365	.671	.014	1.000

224 sample size

± .131 critical value .05 (two-tail)
± .172 critical value .01 (two-tail)

Figure 10. Correlation Matrix Analyses

Table 4. Distance of Districts to Nausori Town

District	Proximity to Nausori (Km)
NAITASIRI	6.2
VUNA	11.2
NAVUAKECE	24.9
NABAITAVO	26.1
WAIDINA	27.7
RARA	28.5
SOLOIRA	30.9

WAIMA	31.6
LUTU	32.2
MATAILOBAU	34.1
NADARAVAKAWALU	39.1
VIRIA	47.3
NAGONENICOLO	48.8
NOIMALU	53.6
MUAIRA	55
NABOBUCO	62.6

4.0   Discussion

Global Aquaculture and Mariculture production retrieved from Literature seen in Table 1 above relays an increasing trend from the 1990-2012 periods. The production steadily increased year after year and was at its highest in 2012 with a contribution of 66,633,253 million tonnes. From this, 38,599 tonnes was finfish and 2,530 tonnes was crustaceans. In 2007, Fiji's overall aquaculture production contributed a total of 247 tonnes valued at less than $3 million dollars.   This decreased in 2012 with a contribution of 0.1% to the total Aquaculture production only 200 tonnes. The decline in 2012s' production is often closely associated to the political instability of 2000. Such instability in Fiji had detrimental economic impacts where declines in the Gross Domestic Product happened straight after such events. These were mostly driven by investors losing confidence due to political instability and the welfare of their investments. "G.D.P decreased from 25% two decades ago to less than 10%" (Wadan Narsey, 2012). For many commodities, declines in production were also evident at the same period.

There were 1390 inshore fishing licenses issued by the Fiji fisheries department in 2014. For the central division of which Naitasiri belongs a total of 331 licenses were issued. To break this down further, the 38 issuances to the Naitasiri province is saw 15 for Marine use and 28 for freshwater. The low number in licenses could be linked directly to the lesser marine fishing grounds belonging to the province. The 28 freshwater licenses accounts for freshwater mussel fishers, fishers who fish waterways and regular gleaners of freshwater prawns and shrimps that is often sold to the local restaurants. The fishing license records for 2014 also tells us of the little opportunities presented by natural circumstances also presented by capture fisheries as there is simply only one marine fishing ground for the whole province situated at Kinoya belonging to the Matanikorovatu clan from Kalabu village. In the past people from the interior or headlands of Naitasiri (almost all districts) often come near the coast to get a taste of the sea and upon visitation approval is first sought from the people of Kalabu. This was the way of the past but today due to modernisation, individualism and legislations on fishing rights the pathway to the easy access to the coast and to fish freely is all but mere history. To fish in this fishing ground today would see the meeting several obligations.

Findings from the research carried out in the Naitasiri province show that there are more farmers situated in the Nabobuco district than in any of the districts. Out of the 65 people sampled in the district the youngest was only 8 years old and the oldest was 77 with the average age 45.2 years old. From this we can infer that most farmers are near the standard retirement age of 55 or most of the famers tend to move into farming at a later stage in life. The existence of the 8 year old farmer could be because of a newer trend of the younger generation being involved or it could simply be a farm passed on to him by an older family member. The mean age for all the farmers in Naitasiri is 45 years and this clarifies the idea that many of this farmers move into aquaculture at a later stage in their life. The lower numbers in younger farmers for the Nabobuco district and for Naitasiri as a whole tells us that younger cohort have been subjected to the rural-urban drift where most people in the rural move to urban and peri-urban areas in search of work, school and for better access to public amenities and services. The whole province farmer status is denoted by 137 active farmers and 87 inactive farmers. The inactive farmers are those that do not have stocked tilapia or prawns in their farms and will not be expecting to get any harvest anytime soon. Active farms in Naitasiri have a mean of 4 years since establishment and the mean farmers age is 43 years. For inactive farmers there was a mean of 6 years of existence since establishment and the mean age for all the inactive farmers was 47 years. Active farmers in the province of Naitasiri inhabit seven districts namely; Lutu, Matailobau, Muaira, Nabaitavo, Nabobuco, Nadaravakawalu and Nagonenicolo. Inactive farmers are located in ten districts namely; Waidina, Viria, Naitasiri, Navuakece, Vuna, Rara, Waima, Soloira, Noimalu and Nagonenicolo. The statistical F-test for means was used to compare active farmers and inactive farmers with proximity to the urban centre Nausori to find out whether distance is the cause of farmers not wanting to farm and losing hope in farming. Results from this test show a p-value of 0.06 where we accept the null hypothesis that there is no significant difference with the two variables and that distance has no major role on whether farms are active or inactive.

The four consecutive years 2012, 2013,2014 and 2015 saw a major jump in the number of new farms being established and the involvement of new farmers. During this four year period there was a total of 104 active farmers which was 61.9% of all the farmers established at this time. Out of all the active farms established during this period, 72% were dug by hand, 27% by machine and the remainder 1% was natural. The major increase in just three years is traceable to the new government's rural development policies for local communities where there has been a shift of focus by the government for people in these rural areas to begin any project on their own, that this was the new criteria to acquiring financial assistance and the fact that the spoon feeding mentality was a thing of the past. Also during this three year period 85% of all the active farms were self financed, 3% was by government and 2% from a group and a National Governmental Organisation. As part of the outreach to rural communities to take up the course and start on their own the government uses the local radio networks to reach out to the remote areas. The move by government has ignited, woken up and has started to change the mindset of people in the rural areas on what should be prioritized in their lives, to participate in national economic growth, adopt new methods, put aside differences as all things are achieved by hard work, utilize idle land that have the potential to be productive and most of all bring in positive change to the people in the respective communities.

Even though change is inevitable as could be identified in this research many rural communities in Fiji like that in Naitasiri still take hold of the norm that males are to lead and women are to provide support. The barrier in gender inequality stills exists in the province where 96% of all the farmers are male and only 4% are female. These findings could also mean that many farmers in Naitasiri are very traditional and probably due to their i-taukei culture see it as a right to lead the farming project. The traditional mindset was when women lead the family this was a sign of disrespect to the men. It is very rare for women to be leading projects in rural Fijian villages but this is slowly changing as there are now three womens group currently involved in aquaculture farming in the Naitasiri province. Two of the four individual farmers are farming at a semi-commercial level and the other two are farming for subsistence use.

Overall constraints to farming in the province includes; 56% request for frys, problems with management 12%, need for technical assistance 12%, flooding which accounts for 6%, accessibility to markets 1%, human resource 2% and feed 3%. For inactive farmers these problems were; heavy downpour destroyed the pond, pond was still under construction and further help was needed, shortage of seed due to the lack of communication with the fisheries department, water source problems, misuse and abuse of project funds by committee members, other commitments, no more interest due to lack of expertise and poor pond design. Many newly established farms were hand dug so there is a high probability that most of the funds will have poor design as they were dug by the villagers. Seed source is highlighted as the major problem faced by the farmers of Naitasiri and this is also similar for many rural communities in Fiji as there is only one government run freshwater hatchery located at Nadurulolou in Nausori. A staggering 84% of all the farmers interviewed indicated that they source their seed from the Nadurulolou government hatchery and most times the government hatchery has not been able to meet their requests. All prawn farmers still depend on the government but tilapia farmers due to the slow response from government have started sourcing frys from the Monasavu Dam, asking neighbours and catching from the wild in their nearby rivers.

To find out the relationship of various variables in freshwater aquaculture the correlation matrix was used and where variables such as age, cost of pond construction, pond area, no. of ponds and Income were compared. Results of this show that only two variables namely; number of ponds and income had a weak negative correlation with age. As age increases the number of ponds decreases and as age increases the income decreases. For a province like Naitasiri this results may be true as most of the farmers manually dug their ponds, self funded their digging and due to the smaller size ponds get less income but this should only be accepted in such a context where technology usage is low. In terms of food security, 98% of all the active farmers in the Naitasiri province eat farmed produce and 2% only farm for selling (commercial farmers). Farmers which are semi-commercial meaning they eat and sell farm produce make up the 37% of the 98% of all farmers that eat what they farm. The reliance on freshwater aquaculture is tremendous for the province. There are five types of common land tenure which include Native Lease, Mataqali owned, Freehold, Crown Lease and other. These will not be elaborated further as there were little problems with land affecting farming identified in the research. The average consumption in a week is 3 days per week for the 98%. Tilapia consists of 86% of all farmed species currently farmed by active farmers, 13% is two or more species farmed together and only 1% of all the farmed species in the province is made up of the freshwater prawn. This tells us that farmers prefer to farm tilapia as it is not too technical when compared to the prawn farming and tilapia species are invasive and opportunistic species that can live and breed almost anywhere in freshwater.

5.0   Conclusion

Global aquaculture steadily rose from 1990 – 2012 but this growth trend was not same for national growth of aquaculture in Fiji, evident with the 2007 production when it rose to 247 tonnes then dropped to 200 tonnes in 2012. Growth trend of various sectors during the same period also endured fluctuations and this at most times were closely linked and associated with the political instability resulting from the coups. The Naitasiri province was issued 38 fishing licenses, 23 of which are for freshwater usage in 2014. The dependency on capture fisheries for the province is low thus the great impetus for more commitment towards freshwater aquaculture. Although the mean age for farming is 45.2 years, age is no longer a barrier and this was seen with an 8 year old boy starting up his own farm. There are indeed older people farming in Naitasiri than young people a revelation of the great impacts of modernisation and urbanization where more young people are prey to the influences promoted by the rural-urban drift. The distance to Nausori have very little to no impact on the status of the farms and this was confirmed by the two sample T-test for means assuming equal variances (p-value = 0.06). The four year period 2012 – 2015 saw a major jump in newly established farms where 72% of all active farms were self dug, 85% of all active farms self financed with only 3% funded by government. The change in mindset towards farming is a result of the new approach by the Fiji government's rural development policies where the spoon feeding approach in assisting farmers is a thing of the past. For farmers to be supported by government they must show the interest and initiative towards farming first and government would meet them half way. There is still gender inequality issues present as the farming is dominated by males accounting for 96% of all the farmers. Constraints to farming are mainly due to the lack of frys, problems with farm management and the very much lacking technical expertise. Others deterrent to farming are the abuse of funds by farmers and the lack of technical expertise to continue farming successfully. Five variables were compared to study the relationships between them, the number of ponds and the income have a weak negative correlation with the Age. This is understandable in the context of this research, there are more old people farming thus the power and stamina during their youthful days have diminished. Farming in the province greatly supports sustenance and livelihoods with 98% of all farmers relying on farmed products for food thus contributing to food security. Tilapia is the most farmed species in the province with prawns being the least farmed, more interest in tilapia is driven by the easy farming requirements, low mortality and aggressiveness to out-compete other species. Freshwater aquaculture in Naitasiri from the results of this study can be confidently said to be improving the nutritional health status for the rural population (protein source), generating much needed income, diversifying activities through aquaculture, supporting food security, allowing idle land to be utilized, contributing to economic growth, improving the socio-economic status and most importantly supporting livelihoods.

6.0   Recommendations

- Gender Inclusion as women have been found to be more caring towards activities concerning family welfare
- Formulate an aquaculture policy to guide farming implementation and end point
- Decentralize Hatchery Operations and establish near farming hotspots
- Government to employ local youth and bring public services to the province to slow down the rural-urban drift phenomenon
- Develop a freshwater aquaculture sector plan for the next five years to give some direction for farming
- More Investment in Digging Equipment to allow for more production
- There needs to be a Feasibility Study before setting up farms
- There needs to be a freshwater farmers cooperative setup to help farmers
- More funding from the government to boost farming
- There needs to be more initiatives focused to increasing Commercial farmers

- New farmers should be identified as soon as possible to ensure maximum production (following the trend observed)
- Subsidize operational costs of Farmers
- Carry out a socio-economic study to find Out the effectiveness of farming for the farmers (Social and Economic Benefits)
- There needs to be further research on Aquaculture in the province (marketing, reproduction and genetic selection, health & disease, economics and Post-harvest management)

7.0 References

Epi Batibasaga. (2015, June). Aquaculture Prsentation. Suva, Fiji.

FAO. (2015). *History of Aquaculture.* Retrieved from www.fao.org: http://www.fao.org/docrep/field/009/ag158e/ag158e02.htm

FAO. (2015). *State of world aquaculture.* Retrieved from Fisheries and Aquaculture Department: http://www.fao.org/fishery/topic/13540/en

FAO. (2005). *Towards Improving Global Information on Aquaculture.* Rome: Food and Agriculture Organisation of the United Nations.

FAO. (2011). *Review of the state of world marine !shery resources.* Rome: FAO Fisheries and Aquaculture Technical Paper No. 569.

FAO. (2012). *The State of the World Fisheries and Aquaculture.* Rome.

FAO. (2015). *INTRODUCTION OF TILAPIA SPECIES AND CONSTRAINTS TO TILAPIA FARMING IN FIJI.* Retrieved from Food and Agriculture Organization: http://www.fao.org/3/contents/a847bb38-345e-547f-96df-969edac7e8cf/AC295E00.htm

FAO. (2015). *Securing Sustainable Small-scale Fisheries.* Retrieved from Food and Agriculture Organization of the United Nations: http://www.fao.org/fishery/topic/16152/154368/en

FAO. (2015). *Statistical systems.* Retrieved from Food and Agriculture Organization of the United Nations: http://www.fao.org/fishery/topic/13410/en

FAO. (2015). *History of Aquaculture.* Retrieved from www.fao.org: http://www.fao.org/docrep/field/009/ag158e/ag158e02.htm

FAO. (2015). *State of world aquaculture.* Retrieved from Fisheries and Aquaculture Department: http://www.fao.org/fishery/topic/13540/en

Foscarini, S. N. (1990). *INTRODUCTION OF TILAPIA SPECIES AND CONSTRAINTS TO TILAPIA FARMING IN FIJI.* Retrieved from FAO.ORG: http://www.fao.org/3/contents/a847bb38-345e-547f-96df-969edac7e8cf/AC295E00.htm

James Teri and Tim Pickering. (2004). *Productivity and constraints in tilapia fish and freshwater prawn aquaculture in Fiji.* Suva: University of the South Pacific.

Meade, J. W. (1989). *Aquaculture Management.* New York: Van Nostrand Reinhold.

Ministry of Fisheries and Forests. (2011). *Annual Report.*

Ministry of Fisheries and Forests. (2011-2013). *Annual Report.*

Ministry of Fisheries and Forests. (2013). *Annual Report.*

Naidu, N. (2015, August 21). Mrs. (E. Batibasaga, Interviewer)

Kudru, M. (2015, May 14). Naitasiri Survey Photos. Suva, Fiji.

Rasuwaki, M. (2015, August 21). Fisheries Assistant. (E. Batibasaga, Interviewer)

SeaWeb. (2004). *At a Crossroads:Will Aquaculture Fulfill the Promise of the Blue Revolution?* Retrieved from A. Seaweb organization Web site: http://www.seaweb.org/resources/documents/reports_crossroads.pdf

SeaWeb. (2004). *At a Crossroads:Will Aquaculture Fulfill the Promise of the Blue Revolution?* Retrieved from A. Seaweb organization Web site: http://www.seaweb.org/resources/documents/reports_crossroads.pdf

SPC. (2005). *Fiji Islands Freshwater Aquaculture Sector Plan.* Noumea: Secretariat of the Pacific Community.

The State of Queensland. (2010-2015). *Types of Aquaculture.* Retrieved from Department of Agriculture and Fisheries: https://www.daf.qld.gov.au/fisheries/aquaculture/overview/types

Wadan Narsey. (2012, March 13). *Our costly coups: the impact on GDP.* Retrieved from wordpress.com: https://narseyonfiji.wordpress.com/2012/03/13/our-costly-coups-the-impact-on-gdp/

8.0 Appendices

8.1 Appendix 1 - Survey Questionnaires

<u>**Freshwater Aquaculture Inventory Survey Form 2015**</u>

Interviewer Name: _______________________ Date: _____________

Farmer: _______________________ M ☐ F ☐ Age ☐

Address: _______________________

Phone: _______________________ Email: _______________________

Locality: _______________________ Village/Province: _______________________

G.P.S point _______________________ Total Pond Area (m²) _____________

No. of Ponds: ☐ Hand Dug: ☐ Machine Dug: ☐

Date of establishment: _____________ Cost of construction ($) _______________________

Source of finance: Self ☐ Govt. ☐ Bank ☐ NGO ☐ Other ☐

Species: Tilapia ☐ Freshwater Prawn ☐ Grass carp ☐ Shrimp ☐ Other ☐

Seed Source: Govt. Hatchery ☐ Other _______________________

Feed Source _______________________

Farm status: Active ☐ Inactive ☐ Reason _______________________

Land Tenure: Freehold ☐ Crown/State Lease ☐ Native Lease ☐ Mataqali ☐
Other ☐

Intensity: Subsistence ☐ Semi-Subsistence ☐ Commercial ☐ Recreation ☐

Production per crop cycle: Kilograms ☐ No of Fish ☐ Fish/Bundle ☐

Average weight at harvest ☐

Market: Roadside ☐ Municipal Markets ☐ Community ☐ Other ☐

Family Consumption per Week _______________________

General Comments
____
____
____
____

8.2 Appendix 2 – Survey Team and Photographs

Coordinator: Epi Qalobula Batibasaga (Student & A/Fisheries Assistant Naitasiri)

Group 1: Peni Vunisau, Teresia[Attachee], Raj Prakash(Driver), Semesa Banuve

Group 2: Manasa Kudru, Kalioni Cagonibure, Aminio Gaunavou, Ro Gasiano(Driver)

All photos below belong to Mr. Manasa Kudru (Kudru, 2015) showing the survey team at work.

8.3 Appendix 3 – Screenshot of aquaculture Inventory in Excel

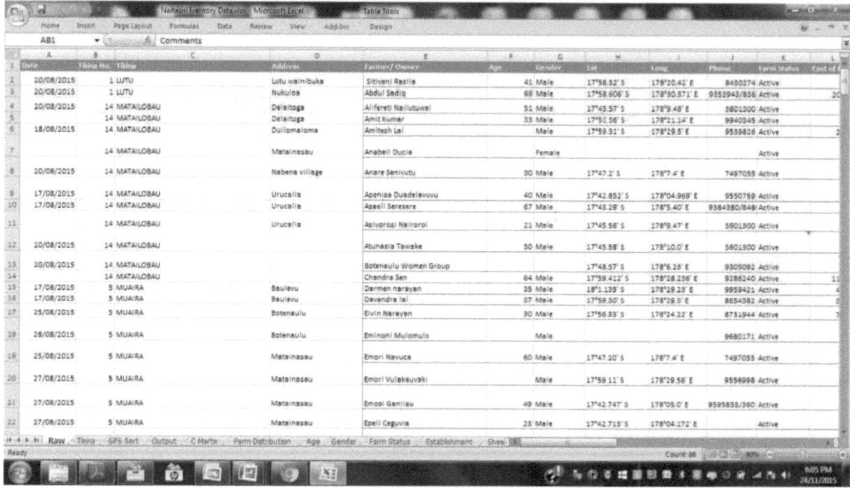

Date	Tikina No.	Tikina	Address	Farmer/Owner	Age	Gender	Lat	Long	Phone	Farm Status	Cost of f
20/08/2015	1	LUTU	Lutu wainibuka	Sitiveni Rasila	41	Male	17°58.52' S	178°20.42' E	8480274	Active	
20/08/2015	1	LUTU	Nukuloa	Abdul Sadiq	69	Male	17°58.606' S	178°30.371' E	9853943/836	Active	20
20/08/2015	14	MATAILOBAU	Delaitoga	Aliferoti Nailutuwai	51	Male	17°45.57' S	178°9.48' E	3601300	Active	
	14	MATAILOBAU	Delaitoga	Amit Kumar	33	Male	17°51.56' S	178°21.14' E	9940945	Active	
18/08/2015	14	MATAILOBAU	Duilomaloma	Amitesh Lal		Male	17°59.91' S	178°29.5' E	9539826	Active	2
	14	MATAILOBAU	Matainasau	Anabell Ducia		Female				Active	
20/08/2015	14	MATAILOBAU	Nabena village	Anare Senivutu	30	Male	17°47.2' S	178°7.4' E	7497055	Active	
17/08/2015	14	MATAILOBAU	Urucalia	Apenisa Duadelavuvu	40	Male	17°42.852' S	178°04.969' E	9550759	Active	
17/08/2015	14	MATAILOBAU	Urucalia	Aseeli Seresere	67	Male	17°45.29' S	178°5.40' E	9364380/846	Active	
	14	MATAILOBAU	Urucalia	Asivorosi Nairoroi	21	Male	17°45.56' S	178°9.47' E	3601300	Active	
20/08/2015	14	MATAILOBAU		Atunasia Tawake	30	Male	17°45.58' S	178°10.0' E	3601300	Active	
30/08/2015	14	MATAILOBAU		Botenaulu Women Group			17°48.57' S	178°6.28' E	9305092	Active	
	14	MATAILOBAU		Chandra Sen	64	Male	17°59.412' S	178°28.236' E	9286240	Active	11
17/08/2015	5	MUAIRA	Baulevu	Darmen narayan	35	Male	18°1.135' S	178°29.23' E	9959421	Active	4
17/08/2015	5	MUAIRA	Baulevu	Devendra lal	87	Male	17°59.30' S	178°29.9' E	8654382	Active	5
25/08/2015	5	MUAIRA	Botenaulu	Elvin Narayan	30	Male	17°56.83' S	178°24.22' E	8751944	Active	7
28/08/2015	5	MUAIRA	Botenaulu	Eminoni Mulomulo		Male				9680171	Active
25/08/2015	5	MUAIRA	Matainasau	Emori Navuca	60	Male	17°47.10' S	178°7.4' E	7497055	Active	
27/08/2015	5	MUAIRA	Matainasau	Emori Vulakauvaki		Male	17°59.11' S	178°29.56' E	9556998	Active	
27/08/2015	5	MUAIRA	Matainasau	Emosi Ganilau	49	Male	17°42.747' S	178°05.0' E	9595833/360	Active	
27/08/2015	5	MUAIRA	Matainasau	Epeli Caguvia	23	Male	17°42.719' S	178°04.172' E		Active	

8.4 Appendix 4 – Freshwater Species Description

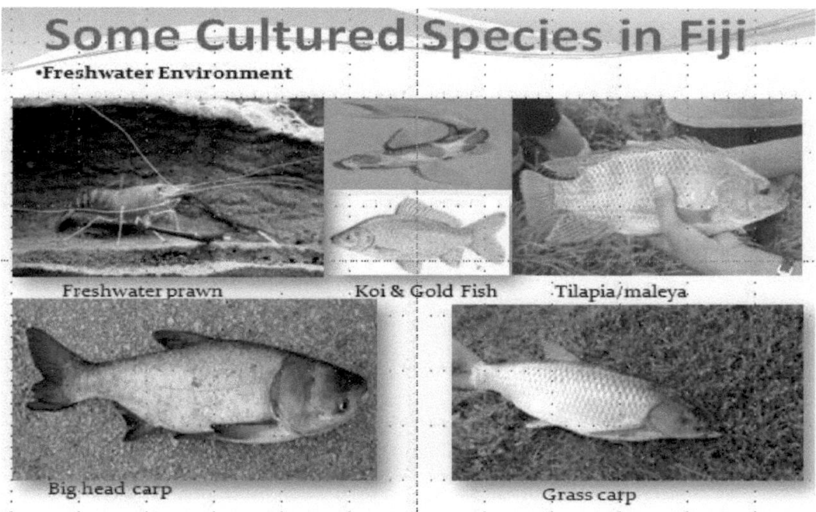

(Epi Batibasaga, 2015) A snapshot taken from Mr. Epi Batibasaga's Presentation

YOUR KNOWLEDGE HAS VALUE

- We will publish your bachelor's and
  master's thesis, essays and papers

- Your own eBook and book -
  sold worldwide in all relevant shops

- Earn money with each sale

Upload your text at www.GRIN.com
and publish for free